TECHNICAL
REPORT

Insights on Aircraft Programmed Depot Maintenance

An Analysis of F-15 PDM

Edward G. Keating, Adam C. Resnick,
Elvira N. Loredo, Richard Hillestad

Prepared for the United States Air Force

The research described in this report was sponsored by the United States Air Force under Contract FA7014-06-C-0001. Further information may be obtained from the Strategic Planning Division, Directorate of Plans, Hq USAF.

Library of Congress Cataloging-in-Publication Data

Insights on aircraft programmed depot maintenance : an analysis of F–15 PDM / Edward G. Keating ... [et al.]
p. cm.
Includes bibliographical references.
ISBN 978-0-8330-4298-9 (pbk. : alk. paper)
1. Eagle (Jet fighter plane)—Maintenance and repair. I. Keating, Edward G. (Edward Geoffrey), 1965–

UG1242.F5I65 2008
358.4'383—dc22

2008021883

The RAND Corporation is a nonprofit research organization providing objective analysis and effective solutions that address the challenges facing the public and private sectors around the world. RAND's publications do not necessarily reflect the opinions of its research clients and sponsors.

Published 2008 by the RAND Corporation
1776 Main Street, P.O. Box 2138, Santa Monica, CA 90407-2138
1200 South Hayes Street, Arlington, VA 22202-5050
4570 Fifth Avenue, Suite 600, Pittsburgh, PA 15213-2665
RAND URL: http://www.rand.org/
To order RAND documents or to obtain additional information, contact
Distribution Services: Telephone: (310) 451-7002;
Fax: (310) 451-6915; Email: order@rand.org

Preface

Lt. Gen. Donald J. Wetekam, then Deputy Chief of Staff for Logistics, Installations, and Mission Support, Headquarters, U.S. Air Force, and Maj. Gen. Arthur B. Morrill III, then Director of Resource Integration, Deputy Chief of Staff for Installations and Mission Support Logistics, asked RAND Project AIR FORCE to develop a series of analyses and models to reveal and explain the effects of changes in Air Force programs on operational capabilities.

As an initial case study, RAND studied the F-15 programmed depot maintenance (PDM) process as it occurs at the Warner Robins Air Logistics Center (WR-ALC) at Robins Air Force Base in central Georgia.

This technical report's primary purpose is to describe WR-ALC's F-15 PDM process as it existed from FY 2004 through FY 2006. We also discuss how it might be expedited, if the Air Force wished to do so.

The research reported here was sponsored by Generals Wetekam and Morrill and was performed as part of an FY 2005–2006 study titled "Capability-Based Programming," which was conducted within the Resource Management Program of RAND Project AIR FORCE. In a report also emanating from this study, Keating and Loredo (2006) presented a methodology to assess a valuation of expedited PDM.

This research is intended to be of interest to Air Force and other Department of Defense maintenance and financial personnel. Along with Keating and Loredo (2006), RAND has conducted numerous studies that touch on related issues. Cook and Graser (2001) studied the effects of lean manufacturing on military airframe acquisition costs, finding a dearth of systematic data collection on the savings being achieved from lean practices. Keating and Camm (2002) noted a general lack of correlation between aircraft flying hours and depot maintenance expenditures. Cook, Ausink, and Roll (2005) urged a rethinking of how the Air Force views sustainment surge. And Loredo, Pyles, and Snyder (2007) discussed PDM capacity assessment.

RAND Project AIR FORCE

RAND Project AIR FORCE (PAF), a division of the RAND Corporation, is the U.S. Air Force's federally funded research and development center for studies and analyses. PAF provides the Air Force with independent analyses of policy alternatives affecting the development, employment, combat readiness, and support of current and future aerospace forces. Research is conducted in four programs: Aerospace Force Development; Manpower, Personnel, and Training; Resource Management; and Strategy and Doctrine.

Additional information about PAF is available on our Web site:
http://www.rand.org/paf/

Contents

Figures

Tables

Summary

This technical report describes the F-15 programmed depot maintenance (PDM) process as it was performed at the Warner Robins Air Logistics Center (WR-ALC) in the FY 2004 through FY 2006 time frame.

The F-15 and Its Programmed Depot Maintenance

The F-15 is an all-weather, extremely maneuverable tactical fighter designed to permit the Air Force to gain and maintain superiority in aerial combat. F-15s are on a six-year PDM cycle, i.e., they are to return for PDM within six years of completion of a visit. WR-ALC has a sequential process that F-15s follow when undergoing PDM. Fuselage and wing work are, however, performed in parallel. (See pp. 4–7.)

F-15 Programmed Depot Maintenance Durations

The mean WR-ALC F-15 PDM visit completed in FY 2006 lasted 119.8 days. This total was down from 130.3 days in FY 2005 but similar to FY 2003 (123.1 days) and FY 2004 (117.5 days) mean durations. (See pp. 9–10.)

In FYs 2002 and 2003, the vast majority of WR-ALC F-15s completed PDM behind schedule. This problem was reduced in recent years, largely because planned durations became more realistic, i.e., longer. (See pp. 10–11.)

In FY 2006, the median F-15 was picked up eight days after WR-ALC completed work. Pickup lags for F-15s based overseas are expected, because they are typically flown overseas in pairs to make more efficient use of aerial tanker refueling. However, even for continental United States (CONUS)–based aircraft, it was not uncommon for operators to wait a week or more to retrieve their completed F-15s. (See pp. 11–13.)

There is considerable variation in how much time aircraft spend at specific steps or cells in the F-15 PDM process. (See pp. 13–14.)

F-15 Programmed Depot Maintenance Part Issues

WR-ALC is concerned about part issues. The PDM line does not have a particularly high priority, so it can wait considerable periods for parts.

One symptom of and adaptation to part problems is "traveling work," i.e., having an aircraft move forward through WR-ALC's cellular flow without all the tasks prescribed in a cell being completed. When the missing part is obtained, the part "catches up" with the aircraft and is installed. (See pp. 15–17.)

Another symptom of and adaptation to part problems is cannibalization. Aircraft that recently entered PDM can serve as sources of cannibalized parts for aircraft that are scheduled to leave sooner.

WR-ALC data suggest that cannibalization is ubiquitous. Data on 99 aircraft entering use in FY 2004 found that every aircraft in the population lost at least one part to cannibalization; only six did not gain a part through cannibalization. (See pp. 16–18.)

Acknowledgments

We especially thank John Fisher and Chandra Thompson for their roles as our points of contact at Robins Air Force Base. We also thank Goran Bencun, Rena Britt, Steve Brooks, Lt. Col. Alex Cruz-Martinez, Doug Daniels, Debra Deckert, Ellen Griffith, Dale Halligan, Norma Jacobs, Mark Johnson, Wendy Johnston, Beth Langford, Fred Lankford, Alan Mathis, Sergeant Kennita Mathis, Jeff Owens, Lorie Snipes, John Stone, Lester Tennant, Greg Whitlock, and Ken Winslette of WR-ALC. Stone provided useful written comments on an earlier draft of this document. We also thank our site-visit hosts: Michael Butler of the Naval Air Systems Command (NAVAIR) fleet readiness center in Jacksonville, Florida; Wayne Chauncey and Darren Jones of the Marine Maintenance Center (MMC) in Albany, Georgia; and Commander Thomas C. Dowden, Supply Corps, U.S. Navy, of San Diego. Randy Miller of Lockheed Martin provided insight on F-22 depot-level maintenance.

We received helpful reviews of an earlier draft of this work from our colleagues Julie Kim and Patrick Mills.

We received helpful insight, suggestions, and comments on this work from Col. Stephen Sheehy of Beale Air Force Base and Maj. Reginald Godbolt of Air Force Materiel Command headquarters. We also thank our colleagues Laura Baldwin, Susan Bowen, Mary Chenoweth, Cynthia Cook, Christine Eibner, Carol Fan, Susan Gates, Gregory Hildebrandt, Kent Hill, Robert Leonard, Raymond Pyles, the late C. Robert Roll, Roberta Shanman, Lauren Skrabala, Mark Wang, and Obaid Younossi for assistance on this research. Candice Riley provided programming support. Sandra Petitjean assisted in creating the figures, Jane Siegel helped prepare the document, and Lisa Bernard edited it.

Earlier versions of this research were briefed at WR-ALC on January 12, 2005; April 12, 2005; June 15, 2005; and November 16, 2005. Maj. Gen. Arthur Morrill was briefed on April 11, 2005; May 11, 2005; and November 4, 2005.

Of course, the authors alone are responsible for errors that remain in the document.

Abbreviations

AE	avionics equipment
AMREP	Aircraft/Missile Maintenance Production/Compression Report
CONUS	continental United States
E&I	evaluation and inspection
FCF	functional check flight
mod	modification
NAVAIR	Naval Air Systems Command
OCONUS	outside the continental United States
PAF	RAND Project AIR FORCE
PDM	programmed depot maintenance
PDMSS	Programmed Depot Maintenance Scheduling System
REMIS	Reliability and Maintainability Information System
WR-ALC	Warner Robins Air Logistics Center

CHAPTER ONE

Introduction

Many U.S. Air Force aircraft undergo programmed depot maintenance (PDM). Depot maintenance involves challenging work, such as extensive aircraft disassembly, that is not done at aircraft home installations. Instead, it occurs at specialized facilities such as the Warner Robins Air Logistics Center (WR-ALC) at Robins Air Force Base in central Georgia. *Programmed* refers to maintenance that occurs on a schedule rather than in response to a specific aircraft's condition. Intermittent PDM is felt to be essential to keep some types of aircraft operating safely and effectively.

PDM increases capability by extending the lives of aircraft. F-15s return to PDM after six years of operation and can be operated for another six years after a completed PDM visit. The six-year PDM "clock" starts on completion of a PDM visit.

This technical report describes the F-15 PDM process as it was performed at WR-ALC in the FY 2004 through FY 2006 time frame.

Chapter Two provides contextual information on the F-15 and how WR-ALC performs F-15 PDM. Chapter Three presents data on WR-ALC F-15 PDM durations. We note that operating commands have not always retrieved their completed F-15s in a timely manner. Chapter Four discusses F-15 PDM part issues, e.g., WR-ALC's adaptations to apparent part shortages. Chapter Five presents a concluding discussion of data that would be needed to better relate PDM resources to F-15 availability.

CHAPTER TWO

The F-15 and Its Programmed Depot Maintenance

This chapter provides background information on the F-15, then describes the F-15 PDM process at WR-ALC.

The F-15 is an all-weather, extremely maneuverable, tactical fighter designed to permit the Air Force to gain and maintain superiority in aerial combat (USAF, 2007). As shown in Table 2.1, there are five F-15 models. The first four models were designed for air-to-air combat, while the newest, most capable, and most expensive one, the F-15E, combines air-to-air and air-to-ground attack capabilities. Table 2.1's data are from the Reliability and Maintainability Information System (REMIS) and are correct as of the end of September 2006.

All five models have two engines and a distinctive double vertical tail design; Figure 2.1 shows a photograph of an F-15.

Table 2.2 shows the assigned location of the 713 F-15s as of the end of September 2006. Six hundred seventy-five were at operating commands' installations; 38 were at Robins Air Force Base or Kimhae.

WR-ALC provides PDM to all five F-15 models. The five models are all handled on the same PDM line at WR-ALC, albeit with some procedural changes related to the aircraft's configuration differences.

The six-year cycle is an engineering specification designed to ensure that the aircraft is safe to fly. Some F-15s receive waivers to fly slightly longer than six years between PDM visits, either due to scheduling issues in the depot system or due to demand for the aircraft at its operating command. However, in general, F-15s are not to fly beyond their six-year cycle without receiving PDM. PDM is felt to be a necessary step to continuing safe operation of an F-15.

Table 2.1
F-15 Models

Model	Seats	Primary Mission	Number Operating	Acceptance Date of Oldest Aircraft	Acceptance Date of Newest Aircraft
A	1	Air-to-air combat	84	March 5, 1975	April 2, 1981
B	2	Air-to-air combat training	14	March 24, 1976	April 18, 1979
C	1	Air-to-air combat	337	May 24, 1979	October 20, 1989
D	2	Air-to-air combat training	54	June 22, 1979	August 11, 1987
E	2	Air-to-ground attack	224	March 11, 1987	September 28, 2004

SOURCE: USAF (2006).

Figure 2.1
An F-15 Eagle

U.S. Air Force photo by MSgt Val Gempis.
RAND *TR528-2.1*

The Programmed Depot Maintenance Process

F-15 PDM is conducted in several buildings at WR-ALC. Figure 2.2 provides an overhead photograph of relevant portions of Robins Air Force Base. An aircraft is received for PDM at the flight line (labeled "F"). An arriving aircraft undergoes an incoming functional test. Afterward, the aircraft is towed to the fuel pit to have its fuel removed. (WR-ALC refers to fuel removal as "defueling.")

The aircraft's engines are then removed, and inaccessible areas of the engines are inspected with a device called a *borescope*. The aircraft, without its engines, is then towed to the modification, or mod, dock in building 83, where fuselage and assembly work occurs. The aircraft is stripped of most of its components, including flight controls, exposing the fuselage. The fuselage is then towed back to the flight-line area for paint removal. The fuselage then returns to building 83 for removal of its panels and wings. The remaining fuselage is then moved to back to the flight-line area and washed. After washing, the fuselage returns to building 83 for the bulk of PDM activities. Fuselage inspection, maintenance, repair, and part replacement are undertaken as approved by the aircraft's operating command, then the fuselage is reassembled.

In parallel to this fuselage work, wing inspection and repair are undertaken at the wing shop in building 140 (which WR-ALC terms "MAN"). The wings are taken apart, inspected, repaired, and reassembled.

Table 2.2
Assigned Locations of F-15s (as of the End of September 2006)

Location Type	Location	F-15 Model A	B	C	D	E	Total
Operating command	Seymour Johnson AFB, N.C.	0	0	0	0	92	92
	Royal Air Force Lakenheath, UK	0	0	22	2	51	75
	Tyndall AFB, Fla.	0	0	44	25	0	69
	Elmendorf AFB, Alaska	0	0	43	4	20	67
	Eglin AFB, Fla.	0	0	56	5	5	66
	Kadena Air Base, Japan	0	0	53	2	0	55
	Mountain Home AFB, Ida.	0	0	17	1	30	48
	Langley AFB, Va.	0	0	29	4	0	33
	Nellis AFB, Nev.	0	0	16	4	12	32
	Kingsley Field Air National Guard Base, Oreg.	0	6	11	3	0	20
	Lambert-St. Louis Air National Guard Base, Mo.	0	1	18	0	1	20
	New Orleans Naval Air Station Joint Reserve Base, La.	16	1	3	0	0	20
	Jacksonville Air Guard Station, Fla.	18	1	0	0	0	19
	Portland Army Reserve Center, Oreg.	17	2	0	0	0	19
	Hickam AFB, Hawaii	16	2	0	0	0	18
	Otis Air National Guard Base, Mass.	11	1	5	1	0	18
	Davis-Monthan AFB, Ariz.	4	0	0	0	0	4
Depot system	Robins AFB, Ga.	0	0	19	0	13	32
	Kimhae Depot, South Korea[a]	2	0	1	3	0	6
Total		84	14	337	54	224	713

SOURCE: USAF (2006).

[a] Kimhae is a Korean Airlines–operated depot maintenance facility that provides F-15 maintenance to Pacific air forces.

Figure 2.2
F-15 Programmed Depot Maintenance Portion of Robins Air Force Base

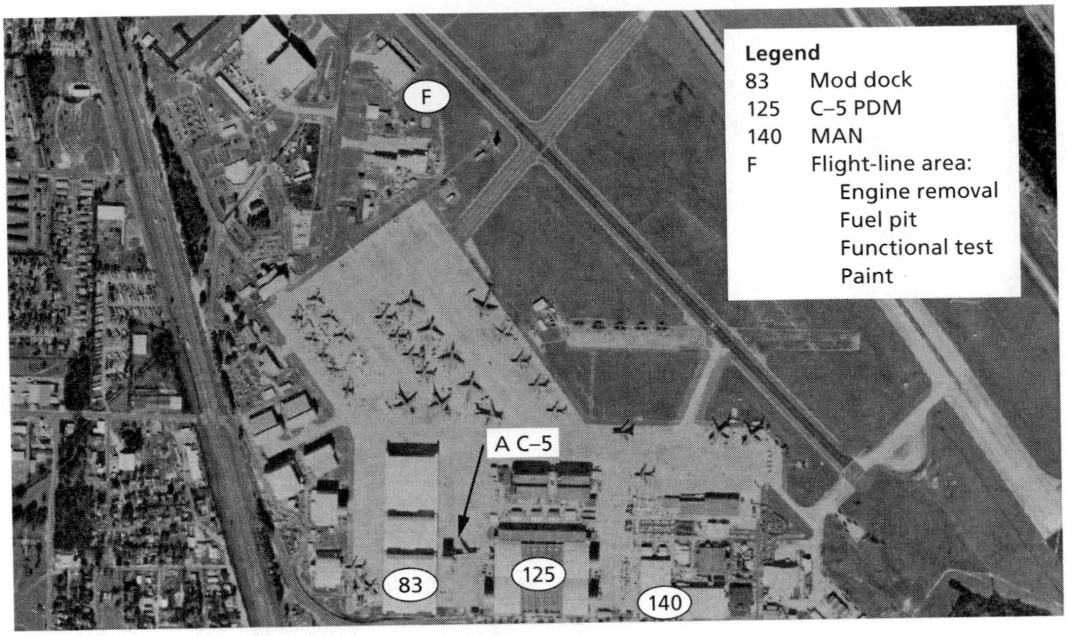

Legend
83 Mod dock
125 C-5 PDM
140 MAN
F Flight-line area:
Engine removal
Fuel pit
Functional test
Paint

F
A C-5
83
125
140

Photo courtesy of U.S. Geological Survey, Department of the Interior.
RAND *TR528-2.2*

When both fuselage work and wing work are completed, the aircraft is reassembled. A series of tests then occurs. The aircraft returns to the flight-line area to check fuel system operations and to ensure that the fuel system is clean. The engines are reinstalled. At the so-called green run, the aircraft's engines and flight controls are operated (while the aircraft is still on the ground). The aircraft then begins functional tests and a functional check flight (FCF). (More than one FCF may be required.) After passing FCF, the aircraft is defueled then painted. Finally the aircraft is refueled and is then stored awaiting pickup by its owner.

A noteworthy aspect of Figure 2.2 is the proximity of WR-ALC's C-5 facility to the F-15's building 83. As shown in this photograph, C-5s sometimes block the entrance to building 83, potentially disrupting WR-ALC's ability to move F-15s in and out of the building.[1]

Table 2.3 lists the steps[2] in the F-15 PDM process. Table 2.3 omits wing-shop work that occurs concurrent to building 83's cells 1 through 3.

Next, we discuss the duration of WR-ALC's F-15 PDM process.

[1] We do not know how common such blocking is. It did, however, occur during a July 2005 RAND visit to WR-ALC. Further, it appeared to be occurring on February 8, 1999, when the U.S. Geological Survey took the photograph shown in Figure 2.2.

[2] *Steps* and *cells* are, for the most part, synonyms in this context. A potential point of confusion, however, is that WR-ALC has made some of its steps into *numbered cells*, while other steps are not so numbered. We use the terms *cells* and *cellular* to refer to all the steps in the PDM process, not simply those that WR-ALC calls "numbered cells."

Table 2.3
F-15 Programmed Depot Maintenance Steps

Area	Description
Receiving	Ground and safety, inventory, egress, incoming operations, drop tank, defuel and purge, preservation
Predock	X-ray, align, remove flight controls, paint removal
Mod dock (building 83)	Cell 1: stripping and evaluation and inspection (E&I)
	Cell 2: vertical work, landing-gear installation, work on E&I write-ups
	Cell 3: tank buildup, installation of flight controls, installation of avionics equipment (AE)
	Cell 4: wing installation
	Cell 5: landing-gear operational check
	Cell 6: flight-control checks
	Cell 7: fuel operational checks
	Cell 8: clean and close panels, install engines
Postdock	Green run: get engines and flight controls working on the ground; functional test
	Wash, etch, alodine
	Paint
	Functional test after paint

SOURCE: Derived by authors during July 2005 visit to WR-ALC.

CHAPTER THREE

F-15 Programmed Depot Maintenance Durations

This chapter discusses WR-ALC's F-15 PDM durations. We start by presenting data on recent years' durations. We then note that WR-ALC's customers have not always picked up completed F-15s quickly. Finally, we present evidence of considerable variability in durations at the PDM cell level.

Figure 3.1 plots the durations (in calendar days) of F-15 PDM visits completed between the beginning of FY 2000 and the end of FY 2006. The horizontal axis displays the date the PDM work was completed. WR-ALC provided these data; they do not include time spent waiting for customers to pick up their aircraft after WR-ALC has completed its tasks. (Customer pickup issues are discussed below.) They do, however, include any delays inducting aircraft into the WR-ALC process.

Figure 3.1
Warner Robins Air Logistics Center's F-15 Programmed Depot Maintenance Durations

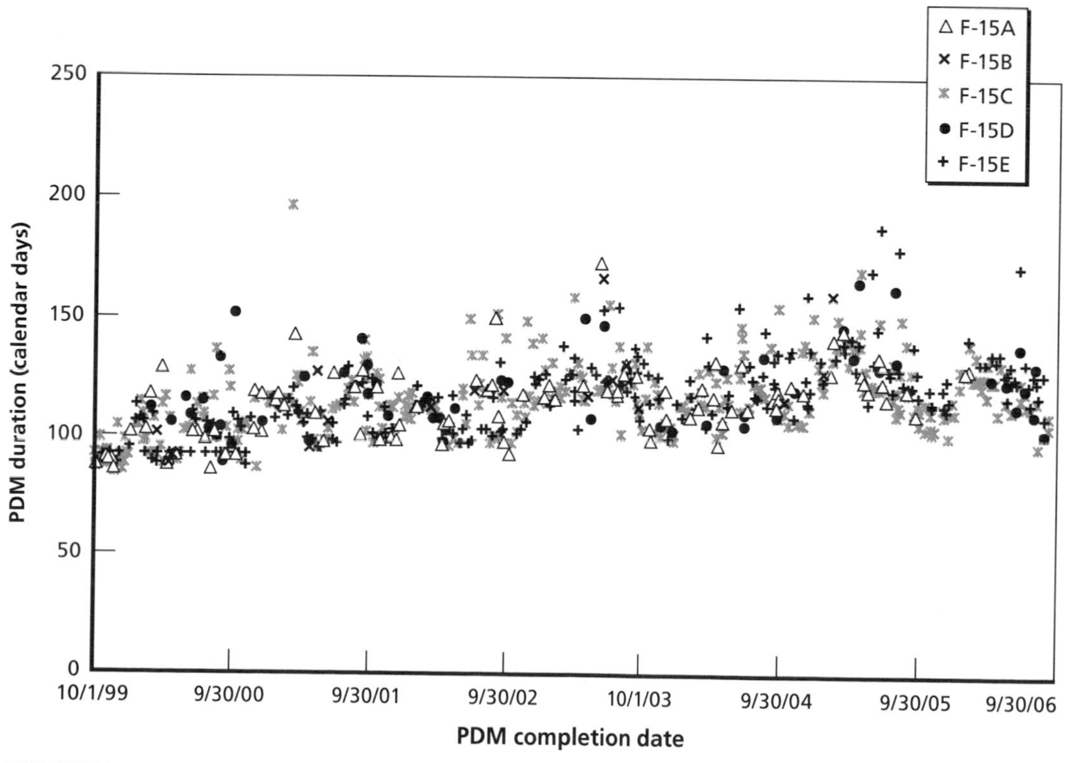

RAND *TR528-3.1*

The average PDM duration was greatest in FY 2005; FY 2006's mean of 119.8 days was more similar to FY 2003 and FY 2004 durations (Table 3.1). WR-ALC produced fewer F-15 PDMs in FY 2006 than in any of the six preceding years. The FYs presented are those in which the PDM work ended. In some cases, the PDM visit commenced in the preceding FY.

We do not fully understand why the long-term trend has been one of increasing PDM durations. One contributing factor is that more is being done to aircraft during PDM, e.g., installation of a Goodrich product called GRID-LOCK® starting in FY 2003. Under this program, honeycombed material in the F-15's flight controls, tails, and wingtips is being replaced. GRID-LOCK technology is designed to be less vulnerable to corrosion and, therefore, less prone to structural failure. (Hebert, 2003, and Lake, 2004, each discussed GRID-LOCK.) Table 3.2 breaks up durations by model.

We do not discern a pattern in durations by model. F-15Es are more complex aircraft than other variants (which one might think would increase maintenance burden), but they are also newer (which one might think would reduce maintenance burden). To a first approximation, WR-ALC's PDM durations do not vary by model.

In Figure 3.2, we plot initially planned PDM durations[1] (the line) against realized PDM durations (the dots). The x-axis value is the date on which WR-ALC inducted the aircraft.

Of the 106 aircraft whose PDM visit started in FY 2002 (between October 1, 2001, and September 30, 2002), 30 (28.3 percent) were completed on or before the initially scheduled completion date. Only 14 of 102 (13.7 percent) aircraft for which PDM commenced in FY 2003 were completed on time or early. Thirty-four of 106 (32.1 percent) FY 2004–inducted aircraft and 51 of 100 (51.0 percent) FY 2005–inducted aircraft were on time or early—but the median initially planned duration for aircraft inducted in FY 2005 was 123.5 days versus 98 days in FY 2002, 100 days in FY 2003, and 111 days in FY 2004.[2] In FY 2004 and, especially, FY 2005, WR-ALC made its planned PDM durations more realistic, i.e., longer.

Table 3.1
Completed Programmed Depot Maintenance Durations

	FY						
Measure	2000	2001	2002	2003	2004	2005	2006
Number completed	116	108	108	102	103	101	87
Mean days	99.5	112.7	111.5	123.1	117.5	130.3	119.8
Standard deviation	11.2	14.8	11.5	13.7	12.2	15.7	11.4
Median days	96.5	112.5	112.0	122.0	116.0	128.0	121.0
75th percentile	106.3	119.3	117.0	127.0	125.0	138.0	126.0
95th percentile	118.5	134.3	133.0	152.9	141.6	162.0	135.0

SOURCE: Britt (2006).

[1] As aircraft go through the PDM process, their planned departure times are updated. WR-ALC refers to the updated departure plan as the Aircraft/Missile Maintenance Production/Compression Report (AMREP) date. Figure 3.2, however, presents initially planned PDM durations, preceding AMREP updates.

[2] The mean initially planned durations were 101.1 days in FY 2002, 103.7 days in FY 2003, 114.5 days in FY 2004, and 125.1 days in FY 2005.

Table 3.2
Completed Programmed Depot Maintenance Durations, by Model

		FY						
F-15 Model	**Measure**	**2000**	**2001**	**2002**	**2003**	**2004**	**2005**	**2006**
A	Completions	18	15	17	13	18	14	2
	Mean	97.2	113.9	112.4	123.3	113.8	125.6	128.5
	Median	92.0	115.0	109.0	120.0	113.0	123.5	128.5
B	Completions	3	3	3	5	—	3	—
	Mean	93.3	109.3	116.7	129.8		133.7	
	Median	91.0	106.0	119.0	124.0		127.0	
C	Completions	50	46	47	47	47	44	46
	Mean	100.5	113.9	113.3	121.6	116.9	127.8	116.1
	Median	98.0	113.0	112.0	120.0	114.0	127.5	115.0
D	Completions	12	8	10	4	8	7	8
	Mean	106.2	124.6	112.1	132.0	112.6	141.1	120.3
	Median	105.0	126.0	110.5	135.0	107.5	134.0	122.0
E	Completions	33	36	31	33	30	33	31
	Mean	97.5	108.3	107.6	123.1	122.2	133.0	124.5
	Median	92.0	109.0	103.0	124.0	122.0	128.0	123.0

SOURCE: Britt (2006).

Figures 3.1 and 3.2 and Tables 3.1 and 3.2 do not include time spent waiting for customers to retrieve completed aircraft. Figure 3.3 presents FY 2006 data on the number of calendar days between when WR completed F-15 PDM and when customers actually retrieved their aircraft. The median duration was pickup eight days after WR-ALC completed work on it.

There were some particularly noticeable lags for aircraft outside the continental United States (OCONUS): We were told that completed F-15s are typically flown overseas paired with another fighter aircraft, an approach that makes more efficient use of aerial tanker refueling. So a completed Lakenheath aircraft might have to wait until another fighter aircraft is ready to cross the Atlantic Ocean.

Even for continental U.S. (CONUS) aircraft, for which air-to-air refueling is not necessary, it was not uncommon for customers to wait a week or more to retrieve their aircraft.

We discussed the issue of delayed aircraft pickup with some F-15 operators. Consistent with the data reported in Figure 3.2, one problem mentioned was that WR-ALC's completion-date projections were not credible. To avoid needlessly sending a pilot to WR-ALC, a customer might not begin the process of arranging a pickup until the aircraft is completely finished by WR-ALC. Pickups cannot occur instantaneously, e.g., pilots may be involved in an exercise or deployment. Another scenario is that a customer may choose to coordinate delivery of an F-15 to WR-ALC for PDM with pickup of an F-15 from PDM, thereby delaying pickup until the next scheduled delivery date. Customers may currently put a greater emphasis on improving the predictability of PDM release dates than on reducing average PDM duration.

Figure 3.2
Planned Versus Actual Programmed Depot Maintenance Duration

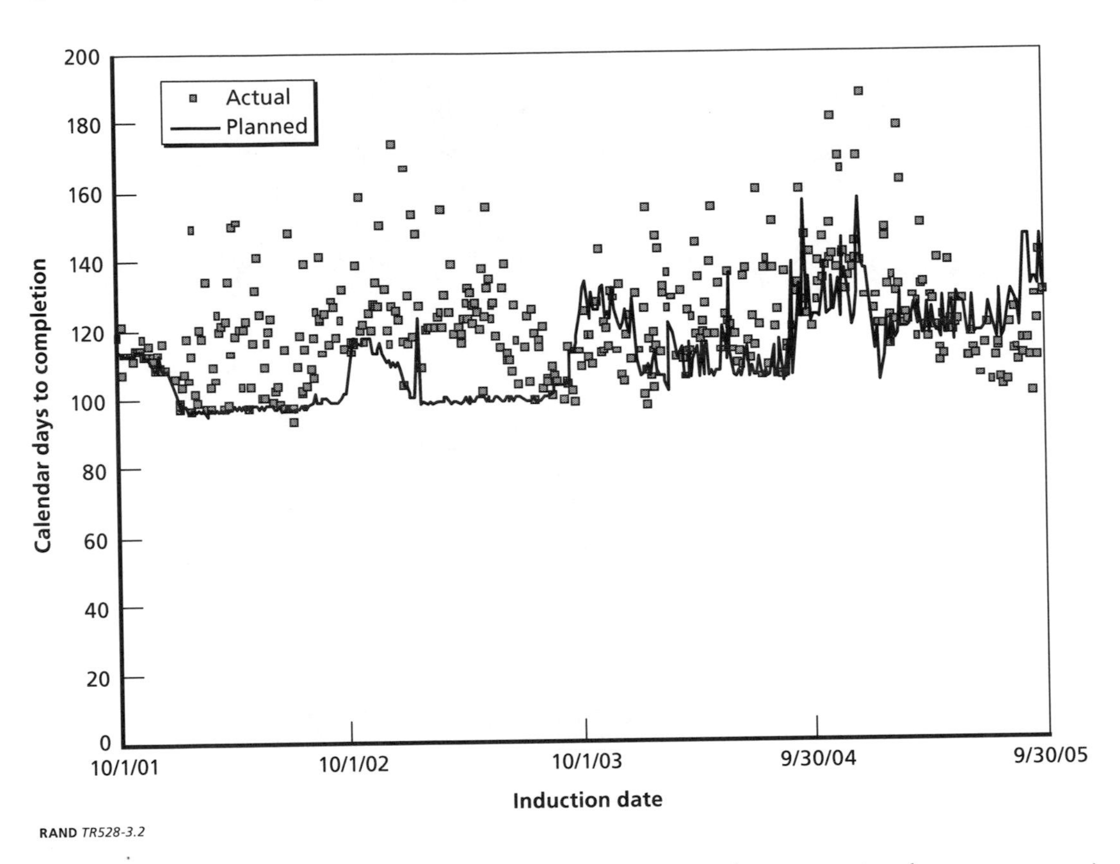

Other inputs, e.g., F-15 pilots, may be the constrained resource in the current environment. If this were true, expediting F-15 aircraft PDM, all by itself, would not have great value.

There may be a short-run versus long-run dichotomy applicable to this issue. In the short run, customers may not be pressing for shorter PDM durations because they have planned their training program around the current system. However, over the long run, F-15 operators could find training and operational benefits associated with having more aircraft available to them.

Cell-Level Duration Variability

As shown in Figure 3.1 and Table 3.1, there is considerable variability in aggregate PDM durations. Duration variance is observed within many F-15 PDM cells. Table 3.3 presents FY 2005 data for about 60 aircraft from the Programmed Depot Maintenance Scheduling System (PDMSS) on the mean and standard deviation of durations, in calendar days, of different steps in the process.

Cell-duration variability can emanate from different causes:

- Aircraft condition: Aircraft may be in inherently better or worse conditions and therefore require less or more time to complete a cell's tasks. (Aircraft condition can vary greatly

Figure 3.3
Days Before FY 2006 Customers Retrieved Their F-15s

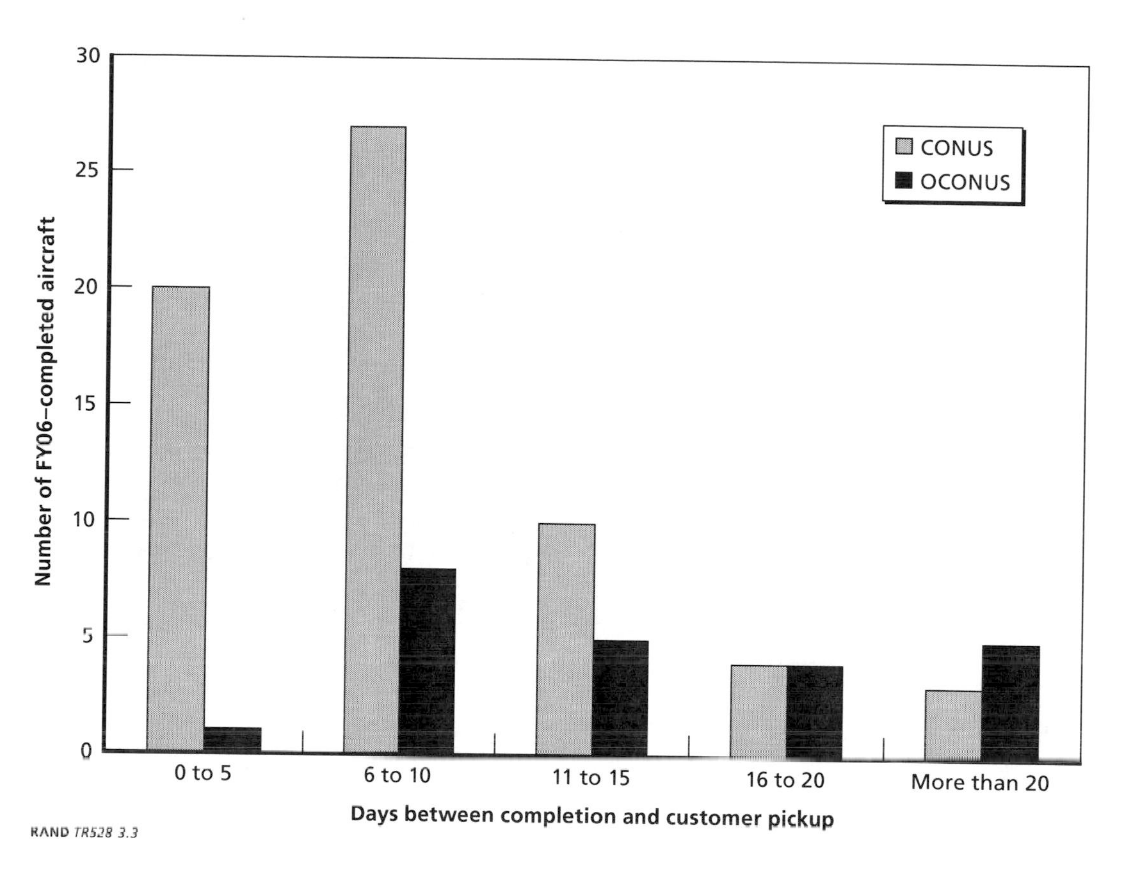

RAND TR528 3.3

Table 3.3
Selected FY 2005 Aircraft Process Distributions

Step	Mean (days)	Standard deviation (days)
Incoming prep and cell 1	15.0	4.2
Cell 2: vertical work	24.1	5.2
Cell 3: tank buildup	11.1	3.4
Cell 4: wing installation	4.4	2.9
Cell 5: landing-gear check	3.0	2.4
Cell 6: flight-control checks	7.4	2.9
Cell 7: fuel operational checks	3.3	1.9
Cell 8: clean and close panels	9.1	4.7
Functional test	13.5	14.4
Paint	3.6	2.1
Functional test after paint	8.4	11.7

SOURCE: Deckert (2005).

based on many factors, including idiosyncrasies in what has happened to the aircraft, the aircraft's age, where and how the aircraft has been operated, and its model.)

- Labor availability: A cell may or may not have sufficient skilled labor to perform its tasks in the planned amount of time.
- Preceding aircraft problems: Cell problems can cascade, e.g., an aircraft may have to wait in cell 4 because an aircraft currently occupying cell 5 is delayed, so the aircraft in cell 4 cannot move forward.
- Part availability: A cell may or may not have the spare parts needed to perform its tasks. If spare-part availability is a problem, the process may be delayed while the cell awaits requisite parts.

Duration variability emanating from aircraft condition variance would be exogenous from WR-ALC's perspective. Aircraft with more problems will likely take longer (unless WR-ALC has slack capacity and "good" aircraft go slower than they could).

The phenomenon of aircraft waiting in a cell because the next cell is fully occupied can cause problem misattribution, as shown in Table 3.3. Problems in cell 6, for instance, might be causing a delay in cell 5. Something is causing the delay—but it may not be in the cell in which the delay is observed.

Next, we discuss part issues in more depth.

CHAPTER FOUR

F-15 Programmed Depot Maintenance Part Issues

This chapter focuses on part issues. In particular, we present evidence of *traveling work* (aircraft moving forward with part holes remaining from earlier cells) and cannibalization. Both are evidence of and adaptations to apparent shortcomings in the F-15 PDM line's ability to get needed parts.

The PDM line does not have a particularly high priority (relative to, say, deployed field-level units), so the PDM line sometimes waits considerable periods for parts.

One symptom of part problems is what WR-ALC terms *traveling work*. A common (and reasonable) adaptation to part problems, we learned, is for an aircraft to move forward through WR-ALC's cellular flow without all the tasks prescribed in a cell being completed, e.g., a requisite part is not yet available. When the part is obtained, the part "catches up" with the aircraft and is installed.

We found evidence consistent with traveling work in the PDMSS record of F-15A 75-019, an aircraft that arrived at WR-ALC on November 1, 2004, and departed on May 10, 2005. While building 83's numbered cells are intended to be sequential, PDMSS indicated considerable overlap in the dates on which different cells' tasks were being accomplished, as shown in Figure 4.1.

In Figure 4.1, each line corresponds to the work of a cell in the PDM process. A cell's line starts the date on which we first see hourly charges to aircraft 75-019 from that cell; it ends on the last date of a charge.

Figure 4.1 indicates that seven cells were "open" (had already made their first charge but had not yet made their last charge to 75-019), for instance, on February 15, 2005. (Cell 1 work had ended, but every other cell listed was open.)

It is also interesting to note that the first hourly charge from cell 8 (incurred December 8, 2004) preceded the first hourly charges in cells 5 (December 9, 2004), 6 (December 20, 2004), and 7 (February 6, 2005).

The PDM path of F-15A 75-019 was not unusual. We analyzed the early FY 2005 PDMSS records of 31 F-15s. Figure 4.2 shows their average first and last charge days in WR-ALC's eight numbered cells measured relative to day 0, the day that REMIS indicates that WR-ALC took possession of the aircraft.

As was true in Figure 4.1, there is considerable overlap in the numbered cells' labor-hour charges in Figure 4.2. Also, consistent with Figure 4.1, cell 8's first hourly charge, on average, preceded cell 5's and cell 7's.

When we showed an earlier version of Figures 4.1 and 4.2 to WR-ALC personnel, they noted how traveling work could explain this result, e.g., earlier cells' work is still being accomplished even while the aircraft resides in a later cell.

Figure 4.1
Evidence of Traveling Work in F-15A 75-019

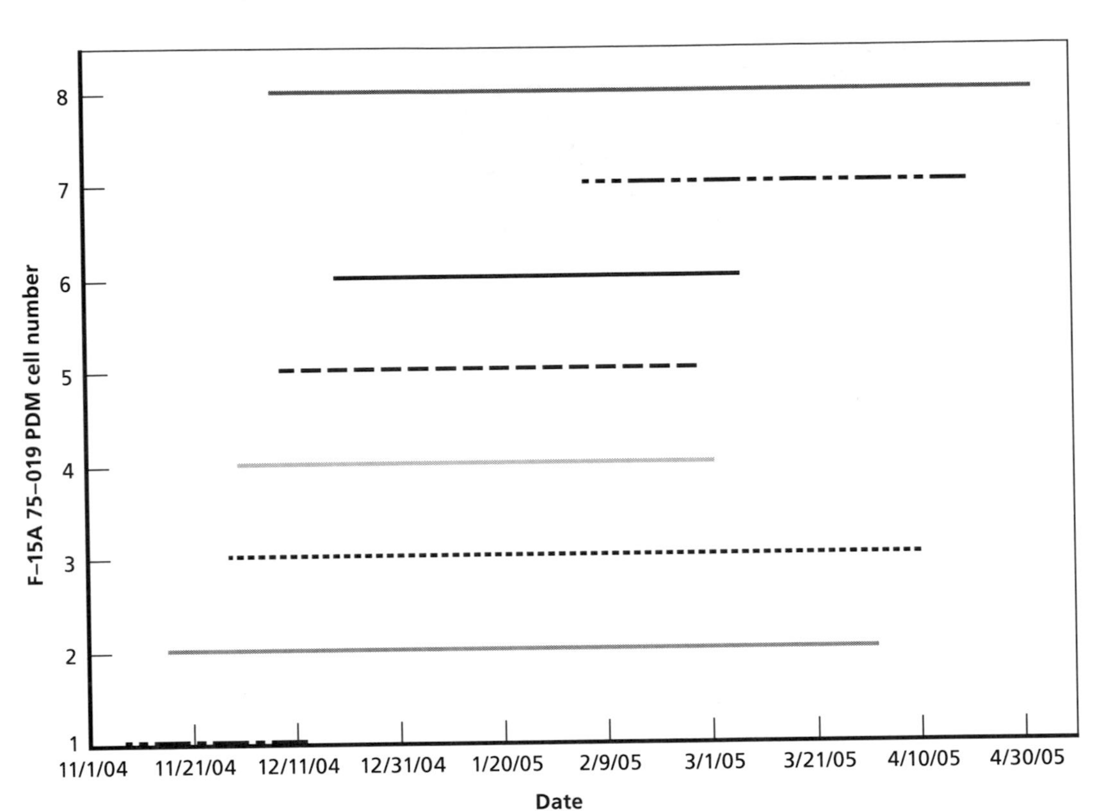

RAND *TR528-4.1*

We were surprised by the extent of apparent cell overlap in these data. Another possible explanation for this finding is imprecision in PDMSS's hour tabulations. But part problem–generated traveling work is clearly a portion of the explanation for Figures 4.1 and 4.2's overlapping horizontal lines.

Another adaptation to a spare-part shortage is to cannibalize requisite parts from other aircraft in lieu of waiting for a part from the supply system. Indeed, WR-ALC's cellular flow is well suited to cannibalization: Aircraft that recently entered PDM can serve as sources of cannibalized parts for aircraft that are scheduled to leave sooner. Indeed, one might imagine multiple generations of cannibalization, e.g., aircraft A takes a part from aircraft B, aircraft B takes the same part from aircraft C, and so forth.

Figure 4.3 plots WR-ALC data on F-15 PDM visits started in FY 2004 and finished by January 28, 2005; there were 99 aircraft with visits fitting this criterion.[1] The x axis shows the number of parts the aircraft lost through cannibalization (typically early in its stay); the y axis shows the number of parts the aircraft gained through cannibalization (typically late in its stay). For any specific aircraft, the number of removals need not equal the number of insertions, as parts could have alternatively come from the regular supply system.

Cannibalization was ubiquitous: Every aircraft in the population lost at least one part to cannibalization; only six of the 99 did not gain a part through cannibalization. The median

---

1 Sixty-five of these aircraft had completed PDM in FY 2004, and 34 did so in FY 2005.

Figure 4.2
Average First and Last Charge Dates, by Warner Robins Air Logistics Center F-15 Numbered Cell

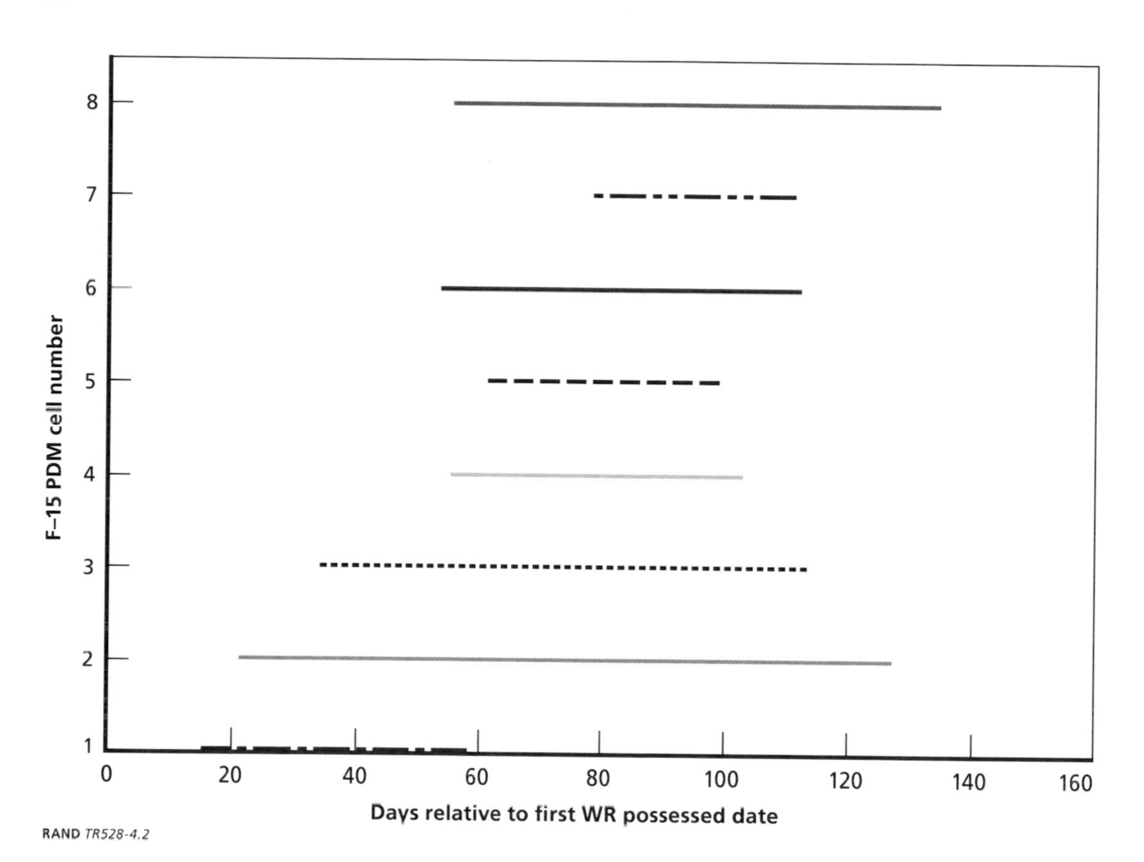

RAND *TR528-4.2*

number of parts both lost and gained was eight (though the parts lost and gained were not necessarily the same).

There are several concerns with cannibalization. Cannibalization can result in parts being broken as they are removed from one aircraft and installed into another. Cannibalization can also result in extra labor through otherwise unnecessary removal and installation of a part. Also, cannibalization creates additional tracking challenges as parts are no longer dedicated to specific aircraft.

These concerns noted, some positive amount of cannibalization is probably optimal. It would be too costly to carry enough parts in inventory to never need cannibalization, and, without large spare-part stockpiles, aircraft would be delayed returning to operating commands without cannibalization. It is an open question whether WR-ALC's current level of F-15 cannibalization is appropriate. One would have to trade off the costs added by cannibalization against the marginal cost of more spare parts needed to reduce cannibalization.

Figure 4.3
FY 2004 F-15 Cannibalization

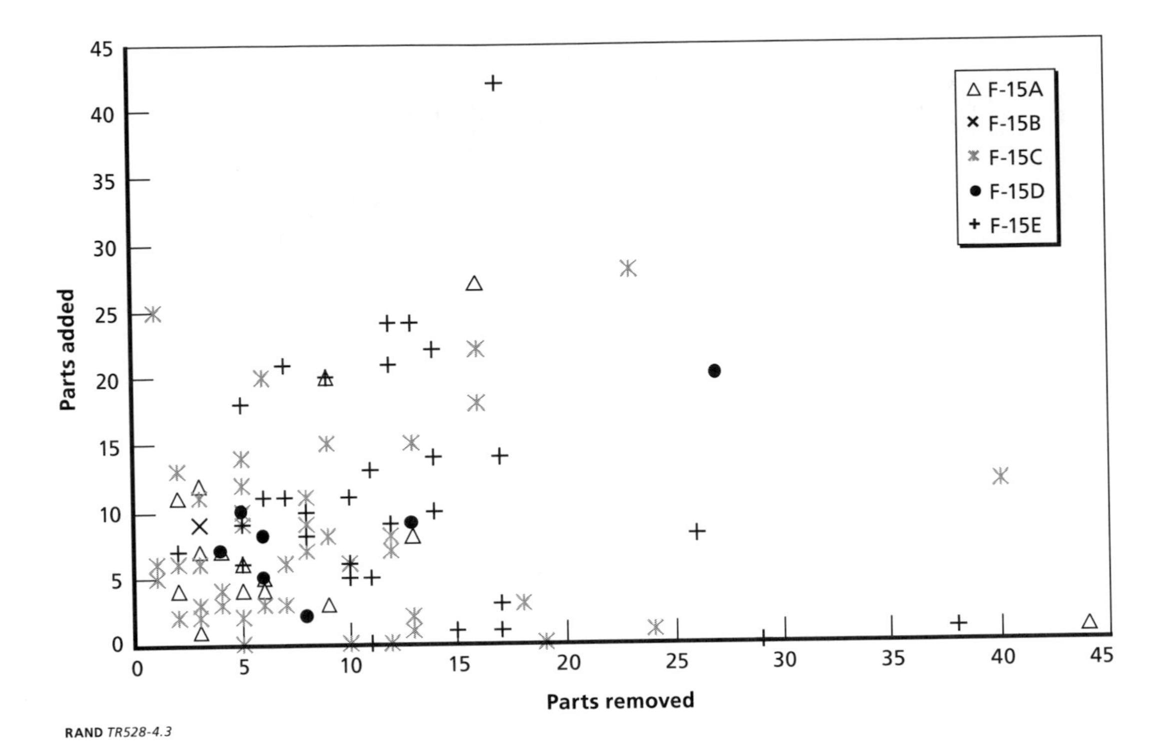

RAND *TR528-4.3*

CHAPTER FIVE

Concluding Discussion

Consider an F-15 progressing through WR-ALC's sequential maintenance process. On the aircraft's arrival at a cell, the cell's personnel need to ascertain what, if anything, needs to be done to the aircraft. Once a diagnosis occurs, personnel, equipment, and supplies need to be available for the required tasks. On completion of those tasks, the aircraft can move forward to the next cell, provided that cell is ready to induct the aircraft.

For a cell's tasks to be completed, necessary equipment, personnel, and supplies must all be present. If equipment is not available, an aircraft will have to wait until, for instance, a different aircraft is no longer using it. If personnel are in short supply, overtime work during nights and weekends may be required. If parts are not readily available, adaptations such as cannibalization and traveling work may occur. Process duration will largely be driven by the worst-performing input; there may be only limited opportunities to, for instance, use labor to substitute for missing supplies (though cannibalization is such a substitution).

If increased resources were made available to the F-15 PDM process, WR-ALC personnel would have to decide where best to allocate those resources. There are different options, including adding equipment, adding labor, and buying additional supplies. The same types of decisions would need to be made if resource levels were cut.

We think that improved data collection could help WR-ALC better allocate depot-level resources. In particular, more and better data might quantify the causes of delays that are occurring in the process. Is skilled labor not available when needed? Are aircraft awaiting parts? Are tooling or equipment levels insufficient at some cells?

To address these questions, one would need detailed data describing the state of each aircraft during its tenure at WR-ALC, what work is being performed on the aircraft, and reasons for any delays that occur. As of our last visit to WR-ALC, it was beginning to track data of this sort with markers and whiteboards posted next to every aircraft.

PDMSS gives some insights on process durations. However, its value could be increased. Several times per day, technicians could note what major task is being performed on each aircraft. WR-ALC organizes work into cells, major tasks, and subtasks. Technicians could collect data at the major task level, recording what major task is being performed on each aircraft at each time increment. PDMSS routinely indicates that an aircraft is receiving work from several cells simultaneously, so it is not currently clear from PDMSS where the aircraft actually was at a point in time. With collected data defining what single major task is being performed at each time increment, WR-ALC could measure task durations more accurately and see when aircraft are idle.

When aircraft are idle, the cause of the delay could be noted. If the causes of delays were tracked and lacking resources were noted (equipment, personnel, supplies), the cost of missing resources could be more firmly established and resources could be better allocated. This data-tabulation exercise could be trivial in some instances, e.g., a data entry might read, "aircraft idle, second shift, no mechanics." However, that information collected across aircraft can provide insight as to the amount of time a typical aircraft spends awaiting maintenance personnel, parts, or other inputs. WR could more easily analyze the trade-off between PDM speed and the levels of key resources, e.g., number of mechanics, the availability of spare parts.

In addition, we recommend more systematic retention and analysis of historical data on aircraft that have completed PDM. Better tabulation of field-reported problems with delivered aircraft would be beneficial as well.

A better, more empirically grounded understanding of the F-15 PDM process could make improvements more feasible. Ultimately, "resource balancing," based on analysis of which cells would most benefit (in terms of reduced process duration) from which additional type of resource (equipment, labor, spare parts) may be possible. Such a calculation might also suggest how PDM resources could be rearranged, e.g., move labor from cell A to cell B to expedite PDM at no or low additional cost. Such calculations could also improve Air Force budgeting by estimating what the effects would be, in terms of process duration, of F-15 PDM budget increases or decreases. At present, there is no straightforward way to translate F-15 PDM resourcing changes into expected changes in PDM durations and, thus, the number of F-15s available to operating commands.

References

Britt, Rena, Warner Robins Air Logistics Center, communication with the authors, November 6, 2006.

Cook, Cynthia R., John A. Ausink, and Charles Robert Roll Jr., *Rethinking How the Air Force Views Sustainment Surge*, Santa Monica, Calif.: RAND Corporation, MG-372-AF, 2005. As of October 17, 2007: http://www.rand.org/pubs/monographs/MG372/

Cook, Cynthia R., and John C. Graser, *Military Airframe Acquisition Costs: The Effects of Lean Manufacturing*, Santa Monica, Calif.: RAND Corporation, MR-1325-AF, 2001. As of October 17, 2007: http://www.rand.org/pubs/monograph_reports/MR1325/

Deckert, Debra, Warner Robins Air Logistics Center, communication with and PDMSS data provided to the authors, 2005.

Hebert, Adam J., "When Aircraft Get Old," *Air Force Magazine: Journal of the Air Force Association*, Vol. 86, No. 1, January 2003. As of February 13, 2006: http://www.afa.org/magazine/jan2003/0103aircraft.html

Keating, Edward G., and Frank Camm, *How Should the U.S. Air Force Depot Maintenance Activity Group Be Funded? Insights from Expenditure and Flying Hour Data*, Santa Monica, Calif.: RAND Corporation, MR-1487-AF, 2002. As of October 17, 2007: http://www.rand.org/pubs/monograph_reports/MR1487/

Keating, Edward G., and Elvira N. Loredo, *Valuing Programmed Depot Maintenance Speed: An Analysis of F-15 PDM*, Santa Monica, Calif.: RAND Corporation, TR-377-AF, 2006. As of October 16, 2007: http://www.rand.org/pubs/technical_reports/TR377/

Lake, Jon, "Grid-Lock Offers Safety Gain on Ageing F-15s," *Flight Daily News*, July 20, 2004.

Loredo, Elvira N., Raymond A. Pyles, and Don Snyder, *Programmed Depot Maintenance Capacity Assessment Tool: Workloads, Capacity, and Availability*, Santa Monica, Calif.: RAND Corporation, MG-519-AF, 2007. As of October 17, 2007: http://www.rand.org/pubs/monographs/MG519/

Mathis, Sergeant Kennita, Warner Robins Air Logistics Center, aircraft-part cannibalization data provided to the authors, January 2005.

USAF—*see* U.S. Air Force.

U.S. Air Force, "F-15: So Long Sword," *Air Force Link*, undated Web page. As of October 16, 2007: http://www.af.mil/photos/index.asp?galleryID=9&page=16

———, Reliability and Maintainability Information System, October 2006.

———, "F-15 Eagle," *Air Force Link*, October 2007. As of February 13, 2006: http://www.af.mil/factsheets/factsheet.asp?fsID=101